Health 128

真菌茶

Fungi Tea

Gunter Pauli

[比]冈特·鲍利 著

[哥伦]凯瑟琳娜·巴赫 绘

章里西 译

上海遠東出版社

特别感谢以下热心人士对童书工作的支持：

匡志强　宋小华　解　东　厉　云　李　婧　庞英元
李　阳　梁婧婧　刘　丹　冯家宝　熊彩虹　罗淑怡
旷　婉　王婧雯　廖清州　王怡然　王　征　邵　杰
陈强林　陈　果　罗　佳　闫　艳　谢　露　张修博
陈梦竹　刘　灿　李　丹　郭　雯　戴　虹

目录

Contents

一只飞鼠听到莫扎特的音乐，停下来抬头张望。音乐声很清楚，可她看不到周围有任何人在演奏，视野里只有一些叶子形状的大蘑菇。一只在附近树枝上休息的北噪鸦跟飞鼠搭话。

“哇，我刚才看到你在树丛间滑翔的英姿了，真优美呀！你也听到对着蘑菇放的音乐了吗？”北噪鸦问道。

A flying squirrel looks up when she hears Mozart music playing. She can hear it clearly but cannot see any musicians. All she sees are some big leaf-like mushrooms. A Siberian jay sitting on a branch nearby starts a conversation:

"Wow, I just saw you gliding through the trees. Beautiful! Can you also hear the music being played to the mushrooms?" Jay asks.

……对着蘑菇放的音乐……

… music being played to the mushrooms …

……蘑菇没有耳朵！

... mushrooms do not have ears!

“我对植物和真菌了解很多，可以明确地告诉你：蘑菇没有耳朵！那么对着蘑菇播放优美的钢琴曲和小提琴曲的意义何在？”飞鼠表示很疑惑。

“我也没见过长耳朵的蘑菇，但声音是由振动产生的，产生这些美妙古典音乐的振动确实打动了我。所以它们肯定对这些蘑菇也有影响。”

“I know a lot about plants and fungi, and can tell you one thing: mushrooms do not have ears! So what is the use of playing the wonderful sounds of the violin and piano to a field of mushrooms?” Squirrel wonders.

“I have never seen mushrooms with ears either, but sound consists of vibrations – and the vibrations of this beautiful classical music certainly affect me. So they must have an effect on them too.”

“你是认真的吗？那你告诉我：这里的这些红色蘑菇喜欢莫扎特胜过摇滚乐吗？”

“确实如此！另外，这些可不是随便什么老品种的蘑菇，它们赤红多汁的菌盖被用来配制一种健康的真菌茶哦。”

“Are you serious? Then tell me: Do these red mushrooms around here prefer Mozart to hard rock?”

“They do! And these are not just any old type of mushrooms. Their red, fruiting bodies are used to make a very healthy fungi tea.”

……这里的蘑菇更喜欢莫扎特……

... mushrooms around here prefer Mozart ...

……白茶、黄茶、红茶、绿茶、黑茶。

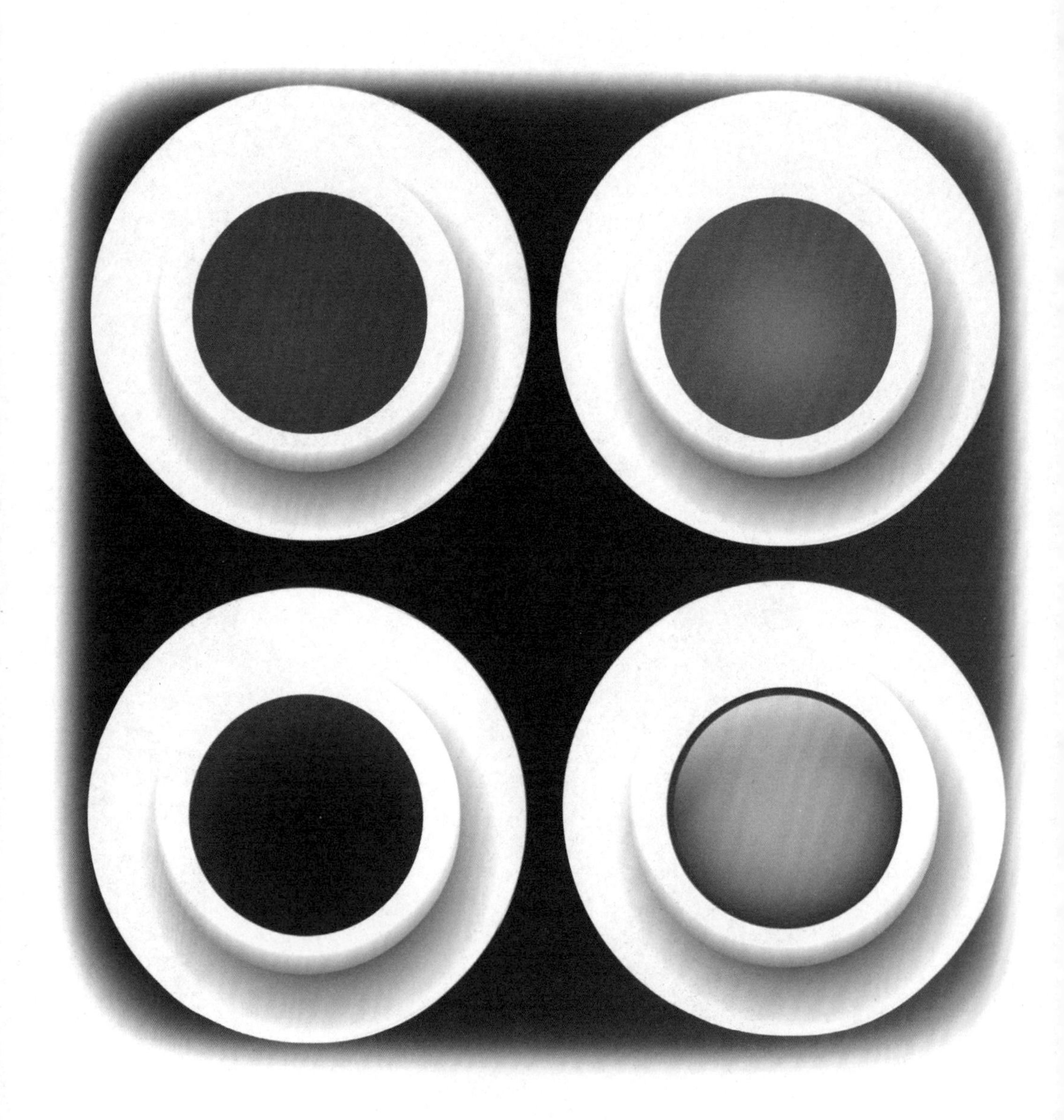

... white, yellow, red, green and black teas.

“嘿，我也很懂茶。我知道白茶、黄茶、红茶、绿茶、黑茶，还知道乌龙茶、草药茶甚至马黛茶，但这是第一次听说真菌茶。”

“世界上总有新知识值得学习。你以前没听说过真菌茶，不代表它不存在……”

“Look, I know a lot about tea too. There are white, yellow, red, green and black teas. There is even oolong, herbal and mate tea, but this is the first time I hear of fungi tea.”

“There is always something new to learn. Just because you have not heard of fungi tea before, doesn’t mean it doesn’t exist ...”

“那怎么制作你说的真菌茶呢？”飞鼠问道。

“你得先找到一株长在死树上的蘑菇，摘下之后泡在热水里来提取里面的营养成分，让香味释放出来。现在你就得到一杯美妙的健康饮品了。”

“So how does one go about making your fungi tea?” Squirrel asks.

“First you need to find a fungus, one that has most likely grown on a dead tree. You break it off and add it to hot water, to extract the nutrients and release those lovely aromas. Now you have a wonderful drink that is very good for your health.”

……泡在热水里……

… add it to hot water …

……另一种类型的药吧！

... a different type of medicine!

“这么看来，我不用茶叶，而是用这些蘑菇也能泡出一杯健康的茶？”

“没错！和用新鲜茶叶或干茶叶泡出的茶不一样，真菌茶可以让你保持健康，如果你生病了，它还能让你更快地痊愈。”

“哦，所以这是一种药了？”

“嗯，算是另一种类型的药吧！”北噪鸦迅速答道。

“So I can make a cup of healthy tea with these mushrooms, instead of tea leaves?”

“Yes! Fungi tea is different from tea made from the fresh or dried first leaves sprouting from a tea bush. It could help your body stay healthy – and get better faster if you were sick.”

“Oh, so it is a medicine?”

“Well, a different type of medicine!” Jay replies quickly.

“你看，药是用来治病、促进身体恢复的，对吧？”

“是，但这些药用蘑菇的特别之处在于可以让你的身体变得更强健，因此让你从一开始就不会得病！你应该多了解一点这些红色、黑色、蓝色、白色、黄色还有紫色的蘑菇茶。”

“Now, medicine is used to cure an illness, and help the body heal when sick, right?”

“Yes, but these medicinal mushrooms are very special in that they makes your body stronger – so that you don’t get sick in the first place! You should discover more about them: these red, black, blue, white, yellow and purple mushroom teas.”

……让你的身体变得更强健……

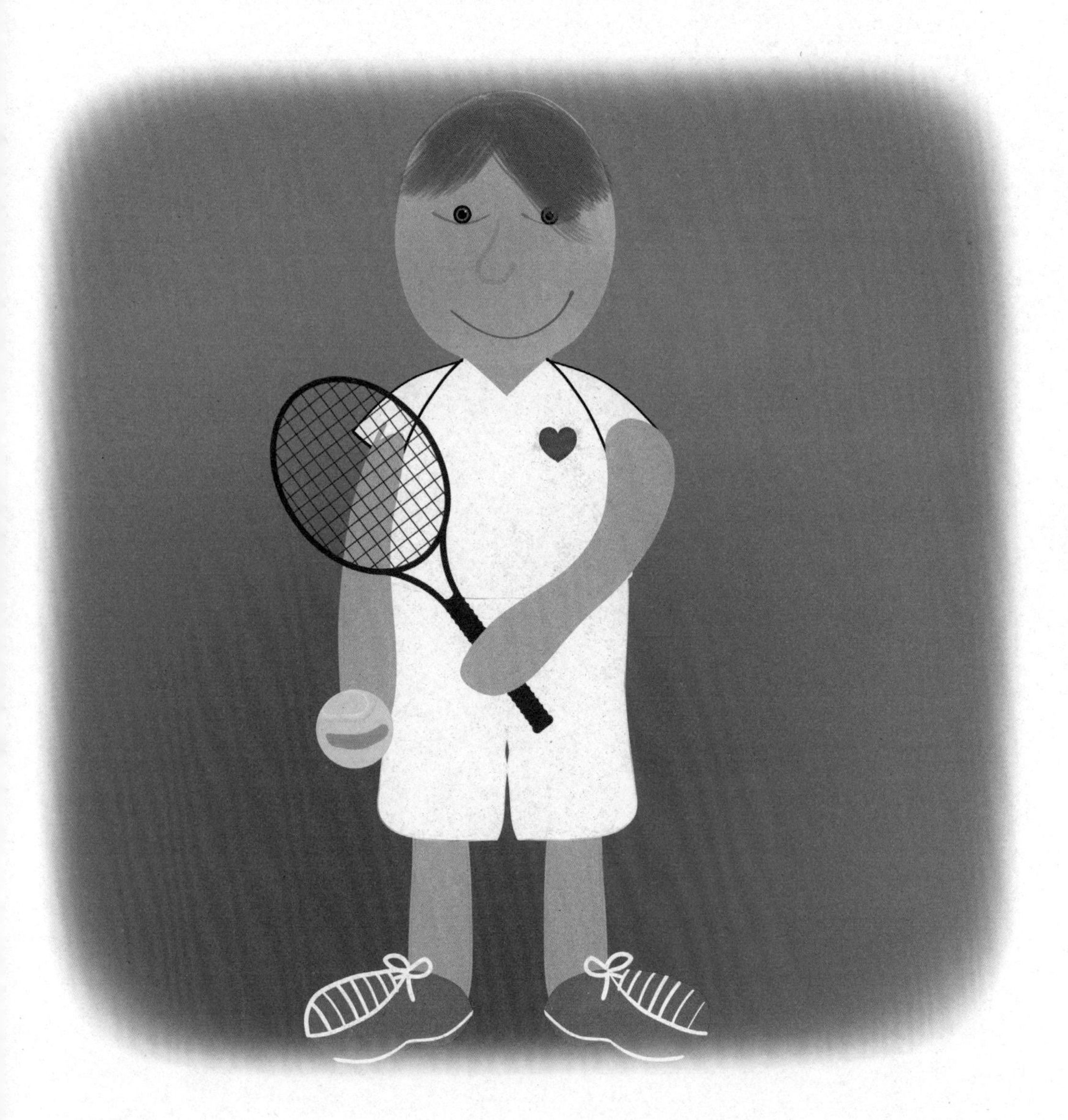

… makes your body stronger …

……只有中国皇帝有权享用呢。

... only the Emperor of China was allowed to have it.

“听起来可以组成一道七色彩虹了！”

北噪鸦点头说道：“红色的那种尤为珍贵，在古代只有中国皇帝有权享用呢。他们会用鹿去寻找这种真菌，就像人们用猪来找乌龟一样。”

“但过去的事情都过去了。”飞鼠说。

“That sounds like they come in all the colours of the rainbow!”

Jay nods and says, “The red one is so special that in the old days only the Emperor of China was allowed to have it. Just like people use pigs to find truffles, they used deer to find it.”

“But the old days are over now,” Squirrel says.

“皇帝的时代的确过去了，但在聪明的科学家的努力下，我们再也不用在森林里寻找这些神奇的药用菌类了。如今真菌已经被广泛种植，每个人都能从中受益。我们一边给真菌放音乐，一边享受它们带来的快乐健康的生活。”

“真是浪漫哪。”

“没错，就像生活在天堂里一样！让咱们听着这美妙的音乐喝了这杯健康的茶，赞美这种能让我们身体强健的药物吧！”

……这仅仅是开始！……

“The time of the Emperors may be gone, but thanks to smart scientists we don’t need to search for these wonderful medicinal mushrooms in the forests anymore. Fungi are now farmed, so everyone can benefit from them – and enjoy a happy and healthy life while we play music for the fungi.”

“That is so romantic.”

“Indeed, it does seem that we are living in paradise! Let’s have a cup of healthy tea, with this beautiful music playing in the background, and celebrate a medicine that strengthens and heals the body.”

... AND IT HAS ONLY JUST BEGUN! ...

……这仅仅是开始！……

... AND IT HAS ONLY JUST BEGUN! ...

Dried mushrooms are tough, and a long soak in hot water is required to release the medicinal molecules from the indigestible chitin.

干蘑菇质地坚硬，需要在热水中长时间浸泡，才能把药用分子从人体无法消化的甲壳质中提炼出来。

Reishi mushroom tea has a very strong and bitter taste. In order to prepare a healthy drink one should not add sugar or honey but rather complement the taste with ginger and/or green tea.

灵芝茶的口感非常苦涩。想要喝到一杯健康的灵芝茶，可以用姜和（或）绿茶调节口感，但不建议加入糖或蜂蜜。

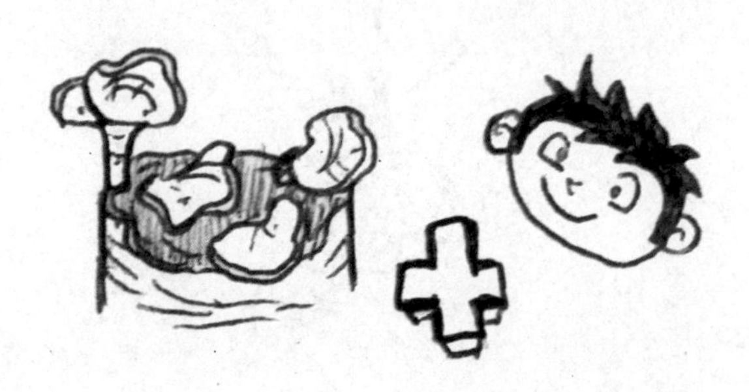

Reishi mushrooms contain triterpenes, which help wounds heal and reduce inflammation. Triterpenes are also found in shark liver and even in human earwax.

灵芝含有的三萜类成分可以促进伤口愈合，减轻炎症反应。人们在鲨鱼的肝脏甚至人类的耳垢中也发现了三萜。

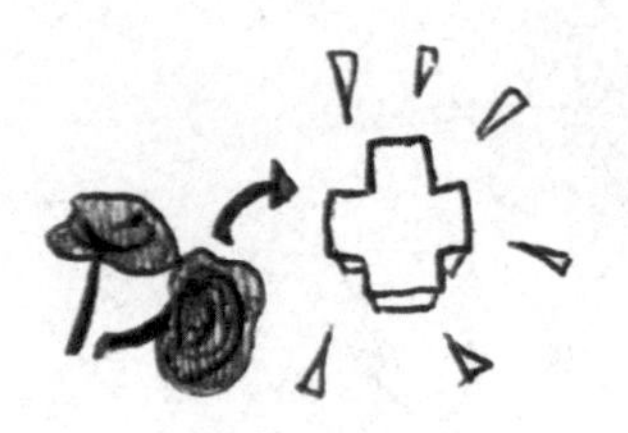

There are six major types of reishi mushrooms: red, black, blue, white, yellow and purple. The red reishi enhances the function of the immune system and vital organs. The black reishi is commonly found in Chinese herbal shops.

灵芝可分为六大类：赤芝、黑芝、青芝、白芝、黄芝及紫芝。赤芝能够增强免疫系统及重要脏器的功能。黑芝在中药店很常见。

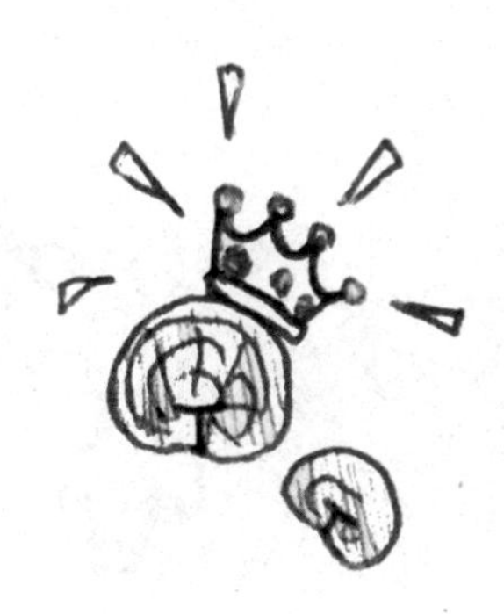

Traditional Chinese Medicine (TCM) considers the red reishi the king of mushrooms. However, TCM also recognises the health benefits of cordyceps, maitake, and shiitake mushrooms.

传统中医认为赤芝乃蘑菇之王，而虫草、灰树花及香菇亦具有保健功效。

The natural red reishi mushroom is found around plum trees, but is extremely rare. It used to be reserved for members of the royal or imperial family only. It is also called mannentake, which (freely translated) means “the one thousand year mushroom”.

野生赤芝通常在李子树附近生长。因其非常罕见，在古代仅供皇室成员享用，故又被称为“万年茸”。

Chaga mushrooms, considered a superfood in Russia, are now available worldwide. Its dark, hard, black exterior looks like charcoal and not at all like a mushroom, and its yellow interior is considered medicinal.

被俄罗斯人认为是超级食物的白桦茸，如今已遍布世界各地。白桦茸外表坚硬、暗黑，看起来像木炭，完全不像蘑菇；其内部呈黄色，被认为有药用价值。

Plants and mushrooms do not hear music, but do perceive sound vibrations. At De Morgenzon wine estate in South Africa, the ripening process of grapes is enhanced with music. Music is also used to promote the growth of good fungi and soil bacteria. The violin seems to have the best effect, and rock music the worst.

植物和菌类并不会聆听音乐，却能感知声音的振动。在南非的朝阳酒庄，人们用音乐来催熟葡萄。音乐同时也用于促进有益真菌及土壤细菌的生长。小提琴曲似乎效果最好，而摇滚乐效果最差。

How is it that we only associate tea with plants and not with fungi as well?

为什么我们说到“茶”时只会想到植物而不是真菌?

What are the differences between Chinese and Western medicine?

中医与西医有何差别?

How many different types of mushrooms do you eat?

你吃过多少种蘑菇?

Do you think that plants and mushrooms will grow better if music is played to them day and night?

如果给植物和蘑菇日夜聆听音乐，你认为它们是否会长得更好?

Do It Yourself!

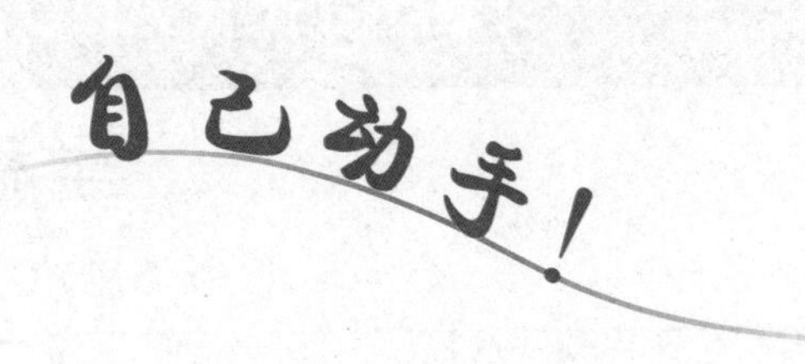

Make a list of all the teas you know of, and the ones you drink regularly. Does your list include any fungi teas? If not, go to a health store and ask for fungi tea. Try different fungi teas and compare their tastes to those of the ones you know. Fungi teas are more like herbal teas, and do not contain any caffeine. Rank the different fungi teas according to their tastes. It may well be that the one with the worst taste is the best for your health! See how the taste correlates with the health benefits and share your findings with your family and friends.

列出你知道的所有茶以及你常喝的几种茶，看看其中是否有真菌茶。如果没有，前往保健品商店咨询真菌茶的相关信息。品尝不同种类的真菌茶，比较一下与你平时喝的茶有何不同。真菌茶的味道更像草本茶，而且不含咖啡因。对各种真菌茶的味道进行排名。最难喝的那款可能对健康反而最有益处！发掘口感和保健功效之间的关系，然后和你的家人、朋友分享你的发现。

学科知识

Academic Knowledge

生物学	灵芝是一种生长在圆木或木屑培养基上的多孔菌；“适应原”是一类能够增强机体抵抗力的物质；灵芝具有“适应原”的作用；飞鼠并不会飞，它们依靠类似“翅膀”的皮肤皱褶在空中滑行。
化　学	菌类中具有活性的药用成分是一种水溶性的多糖，即β-葡聚糖；不同的纺织染料可以用不同的原料制得，如咖啡、红茶和绿茶；制作红茶菌的工艺与制作奶酪类似；甲壳质是一种多糖物质，它参与构成真菌的细胞壁；甲壳质与角蛋白的差异；用酒精可以从灵芝中提取出具有抗炎作用的活性成分。
物　理	真菌茶是通过将菌类浸泡在即将煮沸的水中进行冲泡的；三萜属于甙类，能够降低水溶液表面张力，有助于修复组织损伤。
工程学	红茶菌培养物干燥处理后会变为一种类似于皮革的纺织品，进行模压成形后能够用于生产无缝服装。
经济学	2014年全美红茶菌销量就已达4亿瓶，市场增长预计达到30%；400亿美元规模的茶叶市场中包括了各式各样的茶叶，涵盖了许多有利可图的新市场。
伦理学	西医有部分药品价格过于昂贵以至于目前很少有人能够负担；制药公司经常忽视一些罕见病患者的需求（这些药物的市场规模往往很小）；你不知道某事，并不意味着它不存在。
历　史	红茶菌的产生可追溯到中国古代的秦朝；据说在公元400年前后，一名叫作康普的医生将其引入日本；在明朝李时珍撰写的《本草纲目》中，灵芝第一次被定义为良药；1968年，白桦茸被引入西方世界，其契机是诺贝尔文学奖得主、俄罗斯作家亚历山大·索尔仁尼琴的长篇小说《癌病房》在西方出版。
地　理	白桦茸生长在北半球北纬45°—50°地区，而灵芝大部分生长在热带、亚热带和温带地区。
数　学	在干草堆中找到一根针的概率或是在一片李子树林中发现野生灵芝的概率，均符合数学的概率原理和组合规律。
生活方式	西医的医疗实践基于对器官和症状进行检查后做出的诊断，而中医的目的是将患者身体机能调节至正常，并作为预防医学进行应用；以食为药，是古希腊医学家希波克拉底的观点。
社会学	Kombucha，中文翻译为“茶菌”，日文是“紅茶キノコ”，意为红茶菌；广东话中，茶的发音是“te”；在中国，灵芝代表健康和长寿；在民间传说中，鹿可以嗅出灵芝，如同猪可以找到松露。
心理学	许多东方菌类的药用价值并没有被传统的西医药理学所证实，然而人们相信其效果并乐于使用；信念有助于治疗。
系统论	茶叶渣中仍有一些活性物质残留，不应浪费；红茶菌纤维类似于纤维素，是一种可持续资源，能够用于堆肥。

情感智慧

Emotional Intelligence

北噪鸦

北噪鸦最开始先恭维了飞鼠，接下来并没有直接介绍自己，而是通过向飞鼠提问迅速展开了对话。北噪鸦心直口快，遵循基本逻辑得出了结论并与飞鼠分享。他满怀信心地向飞鼠介绍了一种新产品：真菌茶。他的立场非常鲜明，指出不了解某种事物并不意味着它不存在。北噪鸦通过讲述真菌茶的制备过程进一步阐述了他的观点，然后指出了普通茶与真菌茶的区别。之后他将自己的见解拓展到东西方医学的理念差异。他非常高兴看到现今真菌茶农业技术的发展，不仅富人和皇室成员，所有人都能够享受到真菌茶带给健康的裨益。北噪鸦观察到的这些事实让他自己感到满足。

飞　鼠

飞鼠结束滑翔后四脚落地，对北噪鸦的话迅速给出了务实的回答。起初她直截了当地表明自己不相信北噪鸦说的事情。她提出一个无关紧要的问题来应付，表明她并不太感兴趣，但仍然有礼貌地进行着对话。飞鼠强调，她的茶知识非常丰富，并向北噪鸦进行了展示。飞鼠表示真菌茶对她来说是新鲜事物。当北噪鸦批评飞鼠不关注新鲜事物之后，飞鼠礼貌地询问了如何制作这种茶，并且她认为真菌茶并不是一种茶，而是一种药物。飞鼠一直保持简单、浅显的谈话风格，而北噪鸦则试图进行更深入的交流。

艺术

The Arts

灵芝（真菌类）已经成为中国艺术中非常流行的一个主题。灵芝的菌盖和菌柄非常独特，人们能够非常容易地辨认出来。在中国传统艺术作品中，灵芝旁通常会出现鹿或鹤，因为人们认为鹿能发现灵芝，而鹤和灵芝一样象征长寿。试着用铅笔或毛笔在纸上临摹这些独特的形象，并学习中国艺术的基本元素。

思维拓展

Systems: Making the Connections

真菌在生态系统中的一个重要角色是自然界的清道夫，能够分解、释放植物中蕴含的营养物质，便于动植物对这些营养物质的利用。但它们其实比这做得更多。

真菌的国际市场份额的巨大变化缘于人们发现了大规模商业化种植的方法。真菌培育业创造了大量工作机会，并且形成了可持续的“食物—饲料”循环，在这个过程中能够产生丰富的营养物质。真菌不仅可以食用，越来越多的人还将其制成饮料，如真菌茶。至今没有强力的证据证明真菌具有“完全治愈”某种疾病的药理作用，然而中国和俄罗斯的传统医学通过多年的实践认为，灵芝以及其他菌类可以增强免疫系统功能，促进机体恢复，以及减轻现代西方医学药物的不良反应，比如化疗的不良反应。如今整个社会已经逐渐向更为健康的生活习惯迈进，真菌无疑会在这过程中起到关键作用。我们应该多享用天然的茶类饮料，如灵芝茶，而不是那些富含咖啡因的含糖饮料。幸运的是，真菌培养基的主要成分是废料，如秸秆、玉米芯、咖啡渣及木屑等，这使得我们能够找到生产食品和饮料的新方法，而不用向地球索取更多。只有更好地利用地球慷慨赠予我们的礼物，我们才能做更多事，并变得更健康。

动手能力

Capacity to Implement

让我们开展一项新的业务：销售真菌茶。你认为它的市场前景如何？你家附近是否已经有人开始经营了？如果还没有，那你真是撞上大运了！你并不需要亲自种植真菌，可以选择跟真菌种植户合作，从他们那里购得风干后的真菌，之后只需热水冲泡就能制得受人欢迎的饮料。如果你觉得味道太苦或者太淡，可以尝试加入其他配料，比如生姜或者绿茶。创制你自己的混合配方，仔细思考后和你的朋友、家人与邻居讨论。也许你会比想象中更快地踏入健康产业，提供对人们健康有益的产品！

故事灵感来自

This Fable Is Inspired by

韩省华

Han Shenghua

韩省华在药用菌类种植领域有长达40年的研究和实践经验。在20世纪80年代，他以买卖药用菌类起家，后在张树庭教授的指导下于庆元县远郊农村开展工作。在庆元县他开始试验利用播放古典音乐等新技术在特殊土壤中培育灵芝等菌类。在栽培技术和药用化合物产品方面的丰硕研究成果让他成了备受崇敬的行业领军人物。韩省华除了菌类养殖者这个身份以外，还是一位书画家，在中国各地举办了许多以菌类为主题的画展。韩省华致力于将菌类生物学发扬光大，并邀请了许多海外年轻学者来华，与他们分享他的人生经验。

图书在版编目(CIP)数据

冈特生态童书.第四辑：修订版：全36册：汉英对照 /
(比)冈特·鲍利著；(哥伦)凯瑟琳娜·巴赫绘；
何家振等译.—上海：上海远东出版社，2023
书名原文：Gunter's Fables
ISBN 978-7-5476-1931-5

Ⅰ.①冈… Ⅱ.①冈… ②凯… ③何… Ⅲ.①生态环境-环境保护-儿童读物—汉、英 Ⅳ.①X171.1-49

中国国家版本馆CIP数据核字(2023)第120983号

著作权合同登记号图字09-2023-0612号

策　　划 张　蓉
责任编辑 曹　茜
封面设计 魏　来 李　廉

冈特生态童书
真菌茶
[比]冈特·鲍利　著
[哥伦]凯瑟琳娜·巴赫　绘
章里西　译